CREEPY, KOOKY SCIENCE

Nasty Parasites

Roxanne Troup

Enslow Publishing
101 W. 23rd Street
Suite 240
New York, NY 10011
USA

enslow.com

Published in 2020 by Enslow Publishing, LLC
101 W. 23rd Street, Suite 240, New York, NY 10011

Copyright © 2020 by Enslow Publishing, LLC

All rights reserved.

No part of this book may be reproduced by any means without the written permission of the publisher.

Library of Congress Cataloging-in-Publication Data

Names: Troup, Roxanne, author.
Title: Nasty parasites / Roxanne Troup.
Description: New York : Enslow Publishing, 2020. | Series: Creepy, kooky science | Audience: Grades 5 to 8. | Includes bibliographical references and index.
Identifiers: LCCN 2019009521| ISBN 9781978513815 (library bound) | ISBN 9781978513808 (pbk.).
Subjects: LCSH: Parasites—Juvenile literature. | Parasitism—Juvenile literature.
Classification: LCC QL757 .T76 2020 | DDC 577.8/57—dc23
LC record available at https://lccn.loc.gov/2019009521

Printed in the United States of America

To Our Readers: We have done our best to make sure all websites in this book were active and appropriate when we went to press. However, the author and the publisher have no control over and assume no liability for the material available on those websites or on any websites they may link to. Any comments or suggestions can be sent by email to customerservice@enslow.com.

Photo Credits: Nasty Parasites– Photo research by Bruce Donnola

Cover, pp. 1, 34 Paulo Oliveira/Alamy Stock Photo; p. 5 Louise Gubb/Corbis Historical/Getty Images; p. 6 Robert Pickett/Visuals Unlimited, Inc./Getty Images; p. 8 Darlyne A. Murawski/National Geographic Image Collection/Getty Images; p. 11 Biophoto Associates/PRI/Science Source/Getty Images; p. 13 James H Robinson/Science Source/Getty Images; p. 16 P.F.Mayer/Shutterstock.com; p. 19 José Lino-Neto/Wikimedia Commons/File: Glyptapanteles.png/CC BY 2.5; p. 21 Ian_Redding/iStock/Getty Images; p. 24 Nigel Cattlin/Visuals Unlimited, Inc./Getty Images; p. 26 Unicus/Shutterstock.com; p. 29 mazzzur/iStock/Getty Images; p. 32 Rattiya Thongdumhyu/Shutterstock.com; p. 37 Paulo Oliveira/Alamy Stock Photo; p. 38 Richard Chesher/Science Source/Getty Images

Contents

Introduction	4
1 Nasty Nuisances	7
2 Natural Pest Control	15
3 Killer Plants and Fungi	23
4 Underwater Invaders	31
Chapter Notes	40
Glossary	46
Further Reading	47
Index	48

Introduction

Every living thing needs a habitat. Habitats are like homes. They provide living things with the space and ingredients for survival. Habitats supply food and water. They offer shelter and a safe place to raise young. A habitat can be anywhere. It can be any size. It might be as large as a desert. It might be as small as a leaf. The same habitat can be home to many living things. A habitat can even be another living thing!

Panther chameleons live in the rain forests of Madagascar. They make their home in trees. Trees provide the chameleons with insects to eat, a place to rest and hide, and even dew to drink. The tree is a living thing, but it is also the chameleon's habitat.

Some living things don't look for habitats. They look for hosts. A host is a living thing that has become a habitat for parasites. Parasites steal nutrients from their hosts to survive. A guinea worm is a parasite. It enters human hosts through dirty drinking water. The worm grows by stealing nutrients from the human. Once mature, it burrows out of a person's leg or foot to release its babies.[1] This type of parasite is called an endoparasite. It lives inside the body of its host.

A parasite that lives outside a host is called an ectoparasite. Fleas are ectoparasites. They live on the bodies of furry or feathery animals. A rabbit flea jumps into the soft fur of its host. Its spiky body tangles in the rabbit's coat. The flea bites the rabbit with its needle-like mouth.[2] Then it fills its belly with the rabbit's blood. When the rabbit has babies, the flea jumps onto the babies and lays its eggs. Then it jumps back onto the mother, and the cycle starts again. Now the mother and the babies are both infested with fleas.[3]

There are millions of different kinds of parasites. Some are dangerous to their hosts. They may spread disease or cause terrible pain. Other parasites aren't dangerous. A host might not even realize they are sharing space with a parasite!

Some parasites spend their entire life cycle in the body of one host. Others move from host to host. Some parasites start life in one host, live free for a time, and then find another host where they can mature and reproduce. Parasites are master adapters. When they find a host that meets their needs, they develop

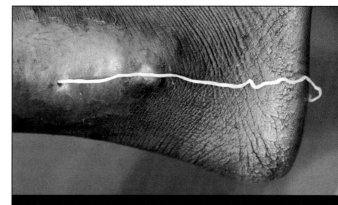

A guinea worm crawls out of a patient's foot in Ghana. Guinea worms were once common in Africa. But thanks to global efforts in providing clean drinking water, the number of infections has dropped significantly.

A tiny cat flea crawls through the hairs of its preferred host. Unless the cat is treated, the flea will hide there for three months feeding and laying up to fifty eggs a day!

specialized ways to stay with that host. They use hooks to hang on tight. They hide by changing shape as they mature. They learn to resist medicines. They may even control host behavior to help with reproduction.

Parasites can be nasty creatures. And they are everywhere. Plants can be parasites. Animals can be, too. Even fungi or bacteria can be parasitic. But not all parasites are bad. As scientists learn more about parasites, they are also discovering ways parasites can help. Gardeners use parasites to control insect populations. Farmers put parasites to work protecting crops. Doctors use parasites to help heal wounds. Experts have even started using parasites to solve ancient mysteries and crimes!

CHAPTER 1

Nasty Nuisances

Parasites have been a part of human history since time began. There are more than four hundred and thirty different parasites that search out human hosts.[1] Some of those parasites go completely unnoticed. Others make humans very sick.

In 1812, French emperor Napoleon Bonaparte (1769–1821) controlled most of Europe. But Napoleon also wanted to control India. The only way into India was through Russia. So that summer, Napoleon ordered his forces to march toward Moscow. On their way, the army ran out of supplies. Men ate what they could find. They hunted. They stole. They caught lice.[2]

Lice are parasites that suck human blood. They can live in hair or on unwashed clothes. Lice also spread diseases such as typhus. Typhus is deadly. It causes high fevers and rashes. It is also highly contagious. Typhus spreads easily in crowded, dirty environments. In the first month of Napoleon's march, eighty

To treat lice, a person has to wash their hair with a special shampoo or lotion, and then comb the lice and their eggs out.

thousand of his men died of typhus and dysentery. Dysentery is another disease caused by parasites. It, too, spreads in crowded, dirty environments.[3] Some scholars estimate that nearly half of Napoleon's army died of parasitic diseases.[4]

Things are a little different now. Though parasites can still be deadly to humans, modern hygiene practices have made most only nasty nuisances.

Itchy Mites

Many parasites that use humans as hosts rely on blood to survive. But human blood is hard to get. Most of it is buried deep within the body. Only a small amount of blood travels close to the surface.

That blood is protected by a thick, flexible shell of skin. How do parasites get to their food source? Some find a way deep into the human body—often through the stomach. Others pierce human skin to expose the tiny blood vessels underneath it.

To pierce skin, parasites need special mouthpieces. These mouthpieces are usually long and sharp. They poke through skin like a needle. When the body recognizes that it has been invaded, it turns on protector cells. These cells release a chemical called histamine. Histamine pushes out invaders and calls for repair cells to patch the break. They also make skin feel itchy.[5]

Mite bites often cause an itchy reaction in humans. Mites are ectoparasites. They are related to spiders. The chigger mite is common in warm, moist climates. It is known for its red body and itchy bites. But adult chiggers don't bite. They live in the soil and eat insects and mosquito eggs. Their babies bite.

Parasitic Crime Fighters

Many parasites are specialized. They live only in specific environments. They only feed on certain hosts. This allows experts to "track" parasites and use them to solve crimes. Once while searching a crime scene in California, investigators were bitten by chiggers. That gave the police an idea. What if the suspect was also bitten? Could that prove his guilt? Police called in the experts. They discovered that the only place chiggers lived in their area was at the crime scene. When police saw bites covering the suspect's body, it proved he'd been there. They sent him to jail.[6]

Chigger larvae are parasites. After they hatch, they climb onto a blade of grass to wait for a passing meal. When someone brushes by, the chigger grabs hold. It looks for a dark, soft spot of skin. The chigger bites. Its spit keeps the bite from healing. The spit turns nearby cells into liquid and makes a sort of blood smoothie for the chigger larvae to drink. Chiggers will feed for four days if undisturbed. Then they fall off. They morph into nymphs (teenage chiggers) and start eating bugs.[7]

Follicle mites are another type of ectoparasite. These little mites are smaller than a grain of salt. They can be seen only with a microscope. Follicle mites live near the roots of people's hair. They eat dead skin and sebum. (Sebum is the oil that makes hair shiny.) Follicle mites can make their home in any skin follicle. A follicle is a tiny hole in the skin. Follicles are also called pores. Follicle mites love the pores on human faces. They especially like the ones at the base of eyelashes and eyebrows. Several mites can fit into one tiny pore. But if too many move in, the hair falls out! Scientists believe that every person on Earth has follicle mites. Thankfully, these tiny creatures don't typically cause problems.[8]

Spreading Fungi

Not all itchy bumps are caused by mites. Some are caused by fungi. Fungi are not plants. Plants make their own food through photosynthesis. Fungi get their food from decomposition. Many people recognize mushrooms and mold as fungi. But yeast (the stuff that makes bread rise) is a type of fungus as well.

Fungi can be parasitic. It can also be food. Some fungus can even be used as medicine. Ringworm is a parasitic fungus that grows on human skin. It has nothing to do with worms. Ringworm starts as a dry, scaly patch of skin. It begins to redden and feel itchy. Then the rash expands, forming its classic circle shape.

Ringworm is a type of mold. It eats keratin. Keratin is the protein that makes hair, skin, and nails. Ringworm likes hot, dark, sweaty keratin. It often appears on feet and around collars. Ringworm is highly contagious. It spreads easily from person to person, or pet to person. And it is difficult to control.

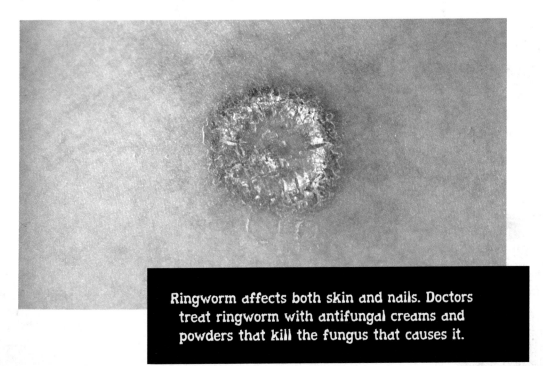

Ringworm affects both skin and nails. Doctors treat ringworm with antifungal creams and powders that kill the fungus that causes it.

Ringworm can even be spread by sharing clothes with someone who is infected.

Ringworm on feet is called athlete's foot. Athlete's foot is very common in locker rooms. It can also show up in public showers like those at a pool. Experts say the best way to control ringworm is to avoid it. Don't share personal belongings with others. Wear cool and dry clothes in summer. And wash hands often, especially after playing with pets.[9]

Growing Worms

True worm parasites are called helminths. (*Helminth* is Greek for "worm.") Scientists group helminths by shape. Some have fat, round bodies like spaghetti. These are called roundworms. Other helminths are flat. They are called flatworms.

A tapeworm is a type of flatworm. Almost every animal, fish, and bird on the planet has a tapeworm that has adapted to its body. The largest is a whale tapeworm. It can grow more than 100 feet (30.5 meters)![10]

Tapeworms have simple bodies. Without a mouth or eyes, their heads work like tiny suction cups. They attach themselves to a host's intestines and start feeding. Tapeworms feed by absorbing nutrients through their skin. When the host eats, small food pieces pass through the stomach and into the intestines. There it mixes with fluids that speed up digestion. Nutrients from food are absorbed into the bloodstream and waste is pushed out to the large intestines.[11] Tapeworms steal some of those nutrients as digested food passes by. As a tapeworm grows, it produces egg sacs. The sacs break off when eggs mature. They are pushed out of the body like any other piece of food waste. Outside the host body, an egg sac dries up. It splits open and releases its eggs. Human tapeworms can live undetected in their hosts for thirty years![12]

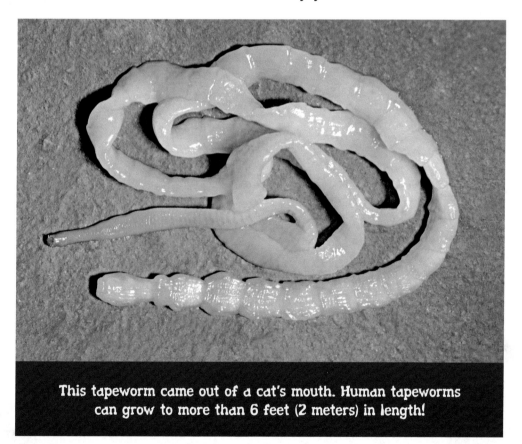

This tapeworm came out of a cat's mouth. Human tapeworms can grow to more than 6 feet (2 meters) in length!

One of the scariest parasitic worms is called a hookworm. Hookworms are tiny roundworms. They have huge mouths … and teeth! Hookworms infect millions of people all over the world. Some people have hundreds of hookworms in their system at once. Hookworms make their hosts feel sick and weak. Hosts may lose weight and even struggle to breathe.[13]

Hookworms enter human hosts through the feet. Walking barefoot through infested soil invites hookworms to burrow through the skin. Once inside, hookworms slice open a blood vein with their teeth and go for a swim. They travel around the body drinking blood. They eventually reach the lungs. There, hookworms make a person cough. The tiny hookworm is coughed up and swallowed. Now it can travel to the intestines and stay awhile. Hookworms can live in the human body for fifteen years![14]

CHAPTER 2

Natural Pest Control

Parasites are important to the ecosystem. They keep plant, animal, and insect populations under control. Many parasites are specialized. They target only one or two hosts. This encourages biodiversity.[1] Biodiversity is the variety of life in an ecosystem. Ecosystems with more variety are healthier. On the island of St. Maarten, two lizards struggle to coexist. The lizards are the same size. They eat the same things. But one lizard is stronger. He has spread around the entire island. The other lizard lives only in the hills. In the hills, the strong lizard is infected with parasites. He grows weak. Now the other lizard has a chance to survive.[2] An ecosystem with several types of lizards is healthier. The variety keeps the food web balanced. If one lizard dies, the other can take its place.

A jewel wasp inspects her prey, a striped cockroach. Soon, she will lead it home and lay one of her eggs on its body. Then, she'll begin hunting another victim.

Parasites encourage biodiversity by making strong hosts weak. They can also control host populations by killing hosts. Many insects begin life as parasites. Larvae use their host's body like an egg. Inside, they are protected. They eat. They grow. Eventually, they "hatch." This kills their host. Insects that go through this cycle and hatch from a host, killing the host, are called parasitoids.[3]

Jewel wasps keep the cockroach population under control. These pretty green wasps attack cockroaches with their stingers. The wasp paralyzes the cockroach. Then it jabs its stinger into the bug's brain. The wasp's venom changes how the cockroach behaves. It can no longer make its own decisions. So the wasp grabs the bug's antenna and leads it home. Then the wasp lays an egg on the cockroach's body. The cockroach waits. It watches the wasp leave. It feels the egg hatch. Still the roach doesn't move. The jewel wasp never returns, but its larva chews a hole in the roach's body. It slips inside and slowly begins to eat. The roach dies. The larva spins a cocoon. In four weeks, it will crawl out of the roach's body and fly away.[4]

Busy Wasps

Farmers and gardeners rely on parasitoids to protect fruit and vegetable crops. The *Aphidius* wasp is a parasitoid. It helps control aphid and mealybug populations. Aphids and mealybugs are tiny bugs that drink plant sap. In the garden, they destroy melon, squash, and tomatoes. In the field, they target potatoes, cotton, and soybeans. These bugs reproduce quickly. If there are too many bugs, plants lose the nutrients they need to produce fruit.

One *Aphidius* wasp can kill two hundred to three hundred mealybugs or aphids during its life![5] An adult wasp lays an egg inside the belly of an aphid or mealybug. The egg hatches. It is

A Piece of the Puzzle

Honeybees are important members of Earth's ecosystem. They pollinate 80 percent of food crops in the United States. They also pollinate the cotton used to make clothes.[6] But honeybee hives are collapsing. Researchers say a parasite is partly to blame. *Varroa destructor* is a small mite. It can fit between the sections of a bee's body. *Varroa* mites enter a hive to lay eggs on bee larvae. The eggs hatch. They feed on the developing larvae. Adult mites also feed on adult bees. They suck their blood and spread disease. Without help, these mites will destroy the hive in a season.[7]

called a larva. The larva lives quietly inside the bug's body for two weeks. It eats. It grows. It spins a cocoon. The cocoon turns the bug into a mummy. Then the adult wasp eats its way out. For two weeks, it will be busy seeking other aphids and mealybugs to host its eggs.

In 2010, scientists released two hundred fifty thousand *Aphidius* wasps in Thailand. Farmers there were dealing with a mealybug outbreak. Mealybugs are not native to Thailand. They are from Paraguay. The bugs were accidently imported on plants shipped into the country. With no natural predators, the mealybugs spread. They began destroying Thailand's cassava crop. Scientists released *Aphidius* wasps hoping to control the mealybug population. It worked![8]

Some wasps lay more than one egg in their hosts. The *Glyptapanteles* wasp can lay up to eighty eggs in one moth caterpillar! The eggs hatch. Wasp larvae fill the caterpillar's body. They begin feeding, but they don't kill the caterpillar—yet. Two weeks later, larvae eat holes in the caterpillar's skin. They wriggle out and begin spinning cocoons. Amazingly, the caterpillar survives. But it doesn't leave. It doesn't eat. It guards the wasp cocoons! When a stinkbug comes near, the caterpillar swings its body from side to side to scare the bug away. The caterpillar stands guard until the wasps emerge from their cocoons. Then it dies.[9]

A caterpillar that's fallen victim to the *Glyptapanteles* wasp makes a great babysitter. It defends the tiny cocoons until they emerge

What makes the caterpillar stay? Why does it guard the wasp cocoons as if they were its own young? Scientists believe the caterpillar becomes a sort of zombie. Zombies are fictional characters from horror movies. They are sometimes called "the undead." Zombies are a strange mix of dead and alive. Their minds are gone. They cannot control their bodies. All they can do is follow orders. And while there is no such thing as human zombies, nature is full of interesting creatures that fit the zombie description.

Amir Grosman, a scientist from Amsterdam, studied the moth caterpillar in Brazil. He wanted to know what was controlling the caterpillar. He noticed that the caterpillar's behavior didn't change until after the wasp larvae "hatched." Grosman wondered if the hatching turned the caterpillar into a zombie. So he copied the hatching wounds on another caterpillar. It didn't behave the same way. Then he dissected a zombie-caterpillar. A few of the *Glyptapanteles* wasp larvae were still living inside! He believes they stayed behind to control the caterpillar. They sacrificed their life to give their siblings a better chance of survival.[10]

Tiny Invaders

Parasitic wasps are small. But some invaders are so tiny they can be seen only by microscope. These tiny invaders don't belong to the animal kingdom. They're not part of the plant kingdom either. They are fungi.

The *Entomophthora muscae* (*E. muscae*) fungus helps control fly populations. "Entomophthora" is a word that means "insect destroyer." *E. muscae* is deadly to flies. It grows in a fly's belly. Then it slowly spreads to the rest of its body. Within a week, the infected fly is dead.

Like most fungus, *E. muscae* spreads by spores. Spores are like tiny seeds. They look like specks of dust. Spores sprout when they land on a fly. The fungus grows into the fly's body. It feeds on the fly's organs and begins to reproduce. Eventually, the fungus sends a signal into the fly's brain. It tells the fly to climb high. The fly obeys. It climbs up a window or blade of grass. It stretches out its body and dies. The fungus continues to grow. Little stalks burst through the fly's skin. They make the fly look

This hoverfly is infected with *E. muscae,* which looks like white growths on its wings and body.

fuzzy. Each stalk contains a spore. In a few hours, the spores will be ripe. They will burst into the air. If they land on another fly, they will grow and start the cycle again.[11]

Another type of fungus targets carpenter ants. *Ophiocordyceps unilateralis (O. unilateralis)* is sometimes called "zombie ant fungus." In damp forests, spores of *O. unilateralis* litter the ground. As a carpenter ant looks for food, it picks up a few spores. The spores sprout. They grow into the ant's body. The fungus spreads and releases chemicals into the ant's brain. The chemicals change the ant's behavior. Instead of climbing back to its nest in the trees, the ant falls to the ground. It searches, but not for food. The ant is looking for the perfect place to die.

    The fungus leads the ant to the underside of a leaf. It is cool there. It stays moist. It is a good place for the fungus to complete its life cycle. *O. unilateralis* directs the ant to bite the leaf. The bite will anchor the ant to the leaf even after it dies. The fungus continues to grow in the dead ant's body. After several days, a stalk shoots out of the ant's head. It swells with spores. Soon it will burst. Spores will shower the forest floor and infect more ants. The "zombie" cycle continues.[12]

CHAPTER 3

Killer Plants and Fungi

Some parasitic fungi target plants instead of insects. In 1845, potato crops in Ireland failed. The leaves withered. The potatoes began to rot in the ground. At harvest, they were mushy and shriveled. They couldn't be eaten. Potato blight had struck. Most people in Ireland lived off potatoes at that time. The poor ate them. The rich sold them. Without potatoes, they couldn't survive. The Potato Famine lasted six years. During that time, nearly one million people died of starvation and disease. Another million left Ireland in search of a better life.[1]

The potato blight is a type of mold. It doesn't directly harm people, but some fungus can. *Claviceps purpurea* is a fungus that infects grain. It poisons the grain and causes a disease called ergot. Ergot makes people sick. It can cause strange visions.

Claviceps purpurea looks like black kernels in infected wheat grains. Farmers help prevent ergot poisoning by rotating crops and cleaning harvested grain using machines.

It can make the body jerk and shake. It can even cause death. Some historians think rye ergot led to the Salem witch trials. In early 1692, young girls in Salem, Massachusetts, began acting strangely. People said they were "bewitched," or under the spell of witches. Others were afraid of being "bewitched," too. They wanted to find the "witches" responsible for the strange behavior. Over the course of a year, two hundred people were accused of practicing magic. They were put on trial or tested. Some were executed. Leaders eventually realized their mistake and stopped the trials.[2] But no one ever discovered the cause of the "bewitching."

Historians think it might have been the bread. The growing season in 1691 was perfect for *Claviceps purpurea* development. Ergot develops in grain during warm, wet summers. If infected grain is made into bread, its toxins spread to people. Diaries from Salem describe symptoms of people in the area. They match the symptoms of ergotism.[3]

Jolly Sapsuckers

Plants can be parasites, too! One famous plant parasite is the mistletoe. In many countries, mistletoe is known as the kissing plant. Each Christmas, people gather wild mistletoe to hang in their homes. The mistletoe is bright and festive. It stays green for weeks after it is cut. And its white berries shine in the light. According to a Victorian tradition, if a man catches a woman under the mistletoe, he can ask for a kiss. Each time

Are Mushrooms Parasites?

Mushrooms are not parasites. They are the fruiting bodies of some fungi. Like apples on a tree, mushrooms are seed carriers. Some fungi that produce mushrooms are parasitic. Parasitic fungi grow on other living things and steal their nutrients. Some fungi give and take nutrients. They are called mycorrhiza. These funguses develop symbiotic relationships with their plant hosts. They help develop strong root systems. Truffles are mycorrhiza. Another group of mushrooms is saprophytes. This group includes most supermarket mushrooms. These mushrooms grow on dead organic matter. They aren't parasites.[4]

26   NASTY PARASITES

Clumps of mistletoe look like green pom-poms in winter. Farmers harvest mistletoe to make extra money. They cut it off fruit trees and sell it as Christmas decorations.

someone is given a kiss, a berry is removed. Once the berries are gone, the kissing must stop.[5]

There are more than a thousand species of mistletoe. They grow wild in forests all over the world. Mistletoe is a hemi-parasite. A hemi-parasite is a plant that gets part of its nutrients from its host. It makes the rest through photosynthesis. Mistletoe grows on tree branches. Its seeds are sticky. Wherever birds drop or poop the seeds out, they stick. But seeds will not sprout on the ground. They must land on a branch. There the seed sends roots into the tree. Mistletoe steals water and nutrients from the branch to start growing. When the first leaves appear, mistletoe begins photosynthesis. But it never makes all its food this way. Mistletoe will continue drawing water from its host. As long as the host lives, mistletoe will survive. It takes five years for mistletoe to produce berries. But that is a good thing. It keeps the plant from destroying too many trees. The only way to rid the host of this invader is to cut off the infected branch.[6]

A different type of mistletoe grows in Australia. But instead of growing as a shrub attached to a tree, this variety grows as a tree. It is called the Australian Christmas tree. Australian Christmas trees steal nutrients from other plants by tapping into their root systems. They do this by growing flesh rings around the roots. Inside each ring is a blade-like device that cuts into the root. The blade is sharp enough to cut through telephone wires and cables buried underground! Each tree taps several root systems. As one of the largest parasites in the world, it can steal sap from plants 361 feet (110 m) away! Some Australian Christmas trees have even been known to tap into their own root systems by mistake.[7]

DNA Pirates

The stemsucker lives in desert areas. It has no roots, no leaves, and no stem. It grows inside the stems of other shrubs. The stemsucker can be seen only when it flowers. Brownish-red buds burst from the stems of infected hosts. They look like tiny berries. Each eruption makes a small round scar on its host. Scientists are still learning about the stemsucker. They believe insects pollinate its flowers. Small animals may spread its seeds. Scientists also think that the stemsucker pirates DNA.[8]

DNA stands for deoxyribonucleic acid. In all living organisms, DNA works like a computer code. It tells cells how to behave. It controls how living things grow and reproduce. It even manages how organisms fight disease and predators. Over the years, some parasites have adapted to stealing their host's DNA. This makes it easier for the parasite to attack. It also makes it harder for the host to resist.

The strangleweed, or dodder, starts life on its own. Its sprouts in moist soil and grows into a leafless vine. As it grows, it searches for a host by its scent.[9] Dodder will stretch up to 1 foot (30 cm) looking for a green plant with tender stems. It twists around its host. It sinks tiny roots into the host's stems. Then dodder hacks the host's DNA! It turns off the plant's self-defense codes.[10] It is now ready to survive on the nutrients from its host. The vine disconnects from the soil. Dodder can live on its own for only five to ten days. But its seeds can survive in the soil for more than twenty years.[11] Once established, the strangleweed vine is nearly impossible to remove.

The master of all DNA pirates is the corpse flower. The corpse flower's proper name is *Rafflesia*. Sometimes it is called a corpse lily or carrion flower. It is the biggest flowering plant in the world. And it is quite possibly the smelliest. Like the stemsucker, the

There are twenty-eight different species of *Rafflesia*. The largest blooms are the size of a kiddie pool. This giant flower is from Indonesia.

corpse flower lives inside the vine of its host. It has no roots or leaves of its own. But this parasite's flower is humungous! It can grow more than 3 feet (1 m) across! It takes six to nine months for a *Rafflesia* flower pod to develop. And when it blooms, it smells like rotten meat.[12] For four or five days, the rubbery flower attracts flies. When flies enter the flower, they are covered in a snot-like substance. The "snot" is the flower's pollen. With the fly's help, *Rafflesia* will soon produce seeds.

Scientists had many questions about *Rafflesia*. How does it grow so big? How are its seeds spread? Does the parasite enter its host through the vine or roots? How does a flower that size not kill its host? Then in 2012, scientists tested *Rafflesia*'s DNA. The data didn't answer their questions, but their findings were shocking. *Rafflesia* appeared to be related to its host! They dug deeper. They discovered that *Rafflesia* had stolen so many DNA pieces, it had made itself invisible! The host plant didn't even know *Rafflesia* was there.[13]

CHAPTER 4

Underwater Invaders

The ocean is Earth's largest ecosystem. It covers nearly 70 percent of the planet.[1] This environment is home to millions of living things. Most are too small to be seen.[2] As these creatures interact, they form food webs and force adaptations. Some adapt into parasites.[3]

Like other parasites, marine parasites feed on the organs and flesh of their hosts. Some even change host behavior. The eye fluke is a worm that lives part of its life in the eye of fish. These parasites start life in the water. They wriggle into fish skin and make their way to the eye. There they hide and grow. When it's time to reproduce, the eye fluke drives the fish closer to the surface. It needs a new host. The fish flips and swims close to the surface. Soon a bird notices. The fish looks like a promising meal.

The bird swoops down. It's an easy catch. But this is exactly where the parasite wants to be. In the bird's stomach, the eye fluke reproduces. It makes eggs that the bird will poop out. When the eggs hit water, they will hatch and the cycle will begin again.[4]

Since most parasitic worms are clear, scientists stain them before examining them under a microscope. This eye fluke is stained pink to help scientists find each of its simple body parts.

Is That My Body?

Some of the strangest parasites in the ocean don't just steal nutrients from their hosts. They become like parts of their host's body! The tongue-eating louse eats and replaces fish tongues. It is an isopod. Isopods have armor-plated bodies and lots of legs. There are three hundred and eighty different species of tongue-eating isopods. Each targets a different host. All tongue-eating lice are born male. They enter hosts through the gills. One turns into a female. It moves into the fish's mouth and sucks blood from its tongue. Eventually the tongue dies and falls out. The female louse finds another place to latch on. To keep eating, the fish uses the louse as its tongue![5]

Sacculina is a copepod. It belongs to the barnacle family and infects crabs. *Sacculina* larvae attach to the hairs on a crab's body. The larva pokes the crab's shell and injects a piece of itself into the crab. This piece

Safe Seafood

Some parasites that infect fish and other seafood can harm humans if eaten. That is why experts recommend cooking seafood before eating it. Seafood should be cooked to 145 degrees Fahrenheit (63 degrees Celsius). Extreme heat kills parasites. Freezing can also kill parasites. Experts say seafood should be frozen for at least a week. They recommend temperatures of -4 degrees Fahrenheit (-20 degrees Celsius). Restaurants that serve fish must follow these guidelines or post warnings on their menus. Some restaurants do both. Sushi restaurants use frozen fish in their dishes and post warnings on their menus.[6]

This clownfish's tongue has been replaced by a tongue-eating louse. The fish can still eat, but the louse may affect how the fish feeds and grows long-term.

grows into a parasite. Soon it will take over the crab's entire body. As *Sacculina* grows, it sends out root-like shoots. The "roots" absorb nutrients from the crab like plant roots absorb nutrients from the ground.[7] *Sacculina* then replaces the crab's reproductive organs. A sac full of *Sacculina* eggs grows out of the crab's belly. The crab thinks the sac is filled with its own eggs. Even male crabs fall for this trick. And male crabs don't produce egg sacs![8] The crab protects the sac and keeps it clean. It cares for the eggs and then releases them as if it were laying its own eggs.

Anelasma is a similar parasite that affects deep-sea sharks. This barnacle attaches to the soft spots on dogfish and lantern sharks. It sends its "roots" under the shark's skin. The "roots" filter nutrients from the blood. And the shark doesn't seem to notice! The open sores around *Anelasma* don't swell or get infected.[9]

Unwelcome Dinner Guests

Jellyfish are some of the most recognized creatures in the ocean. Their shapeless bodies and stinging tentacles are well known. But scientists have mistaken one type of jellyfish for more than two hundred years! Myxozoa are tiny parasitic blobs. They live in freshwater and saltwater environments. They infect fish, reptiles, and even birds. For years, scientists thought this parasite was a protozoan. Then in 2015, tests revealed that it was a tiny jellyfish! It even had its own stinging tentacle.[10]

Myxozoa steal oxygen and nutrients from their host's organs. They also feed on muscles and bones. Myxozoa can cause weight loss and whirling disease.[11] Whirling disease is common in rainbow trout. There is no cure. Myxozoa invade a trout's head and spine. They multiply and twist the bones. They make the fish dizzy. The infected trout swims in wobbly circles. It won't be able to eat or escape predators. Pieces of the dead fish fall into the mud. It decomposes. Myxozoa survive. The parasite waits in the mud for another host to come by. Myxozoa can wait for twenty years.[12]

Another freshwater parasite belongs to the catfish family. Candiru live in the Amazon River. They are less than 2 inches (5 cm) long. They look a bit like toothpicks. Candiru feed on blood. Some people call them toothpick fish. Others call them vampire catfish. Candiru enter the gills of larger fish. They use the sharp spines on their gills to hang on. Then they slice into a host's blood vessel. The candiru feeds on the blood and leaves. It hides in the mud until its meal is digested.[13]

Local legends tell the candiru story differently. They say these tiny invaders are attracted to the smell of urine. That's how they find gill openings. Locals warn visitors not to pee in the water. They say the tiny fish could mistake that smell for a fresh gill. Then it might swim into a human host. Written reports from a hundred years ago tell this story. Of course, scientists don't believe that is

The candiru is a small catfish in the Amazon River that feeds on other fish's blood. Legends surround the fish, but science says the legends have no basis in fact.

true. There is not enough evidence. And while fish gills do release chemical scents similar to human urine, no one knows if candiru find their hosts by smell.[14]

The tiny pearlfish doesn't find its host by smell. It uses feel. Pearlfish live at the bottom of the ocean. They hide (and sometimes feed) in the bodies of sea cucumbers. Sea cucumbers are tube-shaped creatures that eat mud. They don't have eyes. Their mouths face the seafloor. And they breathe through their butts.

NASTY PARASITES

A pearlfish searches for an opening to his new home—inside a sea cucumber. Up to five pearlfish can share space in one sea cucumber!

Sea cucumbers don't have gills. The muscles in their bodies push water in and out of their anal opening. This creates a tiny current. Pearlfish follow the current to find their host. The pearlfish darts inside as the sea cucumber breathes. Parasitic pearlfish feed on its host's organs. Pearlfish that aren't parasites use the sea cucumber's body only to sleep and hide.[15] Clownfish do the same thing with anemones.

There are some crazy parasites in the world! They exist in every environment on Earth. They have adapted to every species as well. But parasites are not monsters. They aren't human enemies. They play important roles in the ecosystem. And as scientists learn more about these tiny invaders, they uncover just how kooky (and sometimes friendly) these "foes" can be.

Chapter Notes

Introduction
1. "Dracunculiasis Eradication: About Guinea-Worm Disease," World Health Organization, https://www.who.int/dracunculiasis/disease/en/.
2. "How Do Fleas Bite and Feed?" FleaScience, https://fleascience.com/flea-encyclopedia/flea-bites/how-do-fleas-bite-and-feed/.
3. Nicola Davies, *What's Eating You? Parasites: The Inside Story* (Cambridge, MA: Candlewick Press, 2007), p. 20.

Chapter 1. Nasty Nuisances
1. Nicola Davies, *What's Eating You? Parasites: The Inside Story* (Cambridge, MA: Candlewick Press, 2007), p. 8.
2. Robert K. D. Peterson, "Section V: Destruction of the Grand Armée," Montana State University, http://www.montana.edu/historybug/napoleon/typhus-russia.html.
3. "Dysentery," MedBroadcast, https://medbroadcast.com/condition/getcondition/dysentery.
4. Peterson.
5. "What Makes Us Itch?" American Academy of Allergy Asthma & Immunology, https://www.aaaai.org/conditions-and-treatments/library/allergy-library/what-makes-us-itch.
6. Rosemary Drisdelle, *Parasites: Tales of Humanity's Most Unwelcome Guests* (Berkeley, CA: University of California Press, 2010), pp. 150–154.

7. Tracy V. Wilson, "How Chiggers Work," HowStuffWorks.com, September 13, 2007, https://animals.howstuffworks.com/arachnids/chigger.htm.
8. Parvaiz Anwar Rather and Iffat Hassan, "Human Demodex Mite: The Versatile Mite of Dermatological Importance," *Indian Journal of Dermatology* 59, no. 1 (January–February 2014): pp. 60–66, https://www.ncbi.nlm.nih.gov/pmc/articles/PMC3884930/.
9. "Ringworm," Mayo Clinic, https://www.mayoclinic.org/diseases-conditions/ringworm-body/symptoms-causes/syc-20353780.
10. Albert Marrin, *Little Monsters: The Creatures That Live on Us and in Us* (New York, NY: Dutton Children's Books, 2011), pp. 109–135.
11. "Your Digestive System & How It Works," National Institute of Diabetes and Digestive and Kidney Diseases, December 2017, https://www.niddk.nih.gov/health-information/digestive-diseases/digestive-system-how-it-works.
12. Marrin, pp. 129–132.
13. Marrin, pp. 115–117.
14. "Hookworm Infections," HealthLine, https://www.healthline.com/health/hookworm#prevention.

Chapter 2. Natural Pest Control

1. Daniel L. Preston and Pieter T. J. Johnson, "Ecological Consequences of Parasitism," Nature Education, 2010, https://www.nature.com/scitable/knowledge/library/ecological-consequences-of-parasitism-13255694.
2. Jos J. Schall, "Parasite-Mediated Competition in Anolis Lizards," *Oecologia* 92, no. 1 (1992): pp. 58–64, https://www.jstor.org/stable/4220127?seq=1#page_scan_tab_contents.
3. Dan Riskin, Ph.D., *Mother Nature Is Trying to Kill You* (New York, NY: Simon & Schuster, 2014), pp. 87–88.

4. Riskin, pp. 87–88.
5. Ker Than, "Parasitic Wasps Unleashed on Insect Pest," Inside Science, September 24, 2014, https://www.insidescience.org/news/parasitic-wasps-unleashed-insect-pest.
6. "Why Bees Are Important," Sustain, https://www.sustainweb.org/foodfacts/bees_are_important/.
7. Ric Bessin, "Varroa Mites Infesting Honeybee Colonies," University of Kentucky Entomology, https://entomology.ca.uky.edu/ef608.
8. Ker Than, "Parasitic Wasp Swarm Unleashed to Fight Pests," National Geographic News, July 19, 2010, https://news.nationalgeographic.com/news/2010/07/100719-parasites-wasps-bugs-cassava-thailand-science-environment/.
9. National Geographic, "Body Invaders," YouTube, April 27, 2009, https://www.youtube.com/watch?v=vMG-LWyNcAs.
10. Ed Yong, "Parasitic Wasp Turns Caterpillars into Head-Banging Bodyguards," *National Geographic*, June 3, 2008, https://www.nationalgeographic.com/science/phenomena/2008/06/03/parasitic-wasp-turns-caterpillars-into-head-banging-bodyguards/.
11. Susan Mahr, "*Entomophthora Muscae*," University of Wisconsin Master Gardener Program, December 4, 2006, https://wimastergardener.org/article/entomophthora-muscae/.
12. Joseph Castro, "Zombie Fungus Enslaves Only Its Favorite Ant Brains," Live Science, September 9, 2014, https://www.livescience.com/47751-zombie-fungus-picky-about-ant-brains.html.

Chapter 3. Killer Plants and Fungi

1. "Irish Potato Famine," History, August 21, 2018, https://www.history.com/topics/immigration/irish-potato-famine.
2. Jess Blumber, "A Brief History of the Salem Witch Trials," *Smithsonian Magazine*, October 23, 2007, https://www.smithsonianmag.com/history/a-brief-history-of-the-salem-witch-trials-175162489/.

3. PBS, "The Witches Curse: Clues and Evidence," *Secrets of the Dead: Unearthing History*, http://www.pbs.org/wnet/secrets/witches-curse-clues-evidence/1501/.
4. "What Is a Mushroom?" Scelta Mushrooms, https://www.sceltamushrooms.com/en/themes/what-is-a-mushroom/.
5. Kate Langrish, "Mistletoe: How Does It Grow and Why Do We Kiss Under It?" *Country Living*, December 20, 2018, https://www.countryliving.com/uk/wildlife/farming/a23006736/mistletoe-farm-kiss/.
6. Brian Barth, "The Enduring Romance of Mistletoe, a Parasite Named After Bird Poop," *Smithsonian Magazine*, December 21, 2017, https://www.smithsonianmag.com/science-nature/poop-tree-parasite-mistletoe-180967621/.
7. Tim Low, "Australia's Giant Parasitic Christmas Tree," *Australian Geographic*, May 15, 2017, https://www.australiangeographic.com.au/blogs/wild-journey/2017/05/australias-giant-parasitic-christmas-tree/.
8. Blackman Lab, "Bloom of the Week: Thurber's Stemsucker," University of California, Berkeley, February 27, 2017, https://nature.berkeley.edu/blackmanlab/Blackman_Lab/Lab_News/Entries/2017/2/27_Bloom_of_the_Week_-_Thurbers_Stemsucker.html.
9. "Dodder Vine Sniffs Out Its Prey," PBS, April 3, 2013, https://www.pbs.org/video/nature-dodder-vine-sniffs-out-its-prey/.
10. Barbara Kennedy, "This Parasitic Plant Hijacks Its Victim's Genes," Futurity, January 21, 2018, https://www.futurity.org/dodder-silences-plant-genes-1659042/.
11. "Dodder," Missouri Botanical Garden, http://www.missouribotanicalgarden.org/gardens-gardening/your-garden/help-for-the-home-gardener/advice-tips-resources/pests-and-problems/weeds/dodder.aspx.
12. Redfern Natural History Productions, "BIGGEST Flower in the World: *Rafflesia Arnoldii*," YouTube, December 19, 2016, https://www.youtube.com/watch?v=0cGRujABwuQ.

13. Jonathan Shaw, "Colossal Blossom: Pursuing the Peculiar Genetics of a Parasitic Plant," *Harvard Magazine*, March–April 2017, https://www.harvardmagazine.com/2017/03/colossal-blossom.

Chapter 4. Underwater Invaders

1. "How Much Water Is There on, in, and above the Earth?" US Geological Survey, December 2, 2016, https://water.usgs.gov/edu/earthhowmuch.html.
2. Jon Copley, "Blue Planet II: Is the Ocean Really the 'Largest Habitat on Earth'?" The Conversation, October 31, 2017, https://theconversation.com/blue-planet-ii-is-the-ocean-really-the-largest-habitat-on-earth-86556.
3. Ed Yong, "Animals Have Evolved into Parasites at Least 200 Times," National Geographic News, July 19, 2016, https://news.nationalgeographic.com/2016/07/animals-evolution-parasites-ed-yong/.
4. Elizabeth Preston, "Parasite Living Inside Fish Eyeball Controls Its Behavior," *New Scientist*, May 4, 2017, https://www.newscientist.com/article/2129880-parasite-living-inside-fish-eyeball-controls-its-behaviour/.
5. Katija Laingui, "K's Kreature Feature: Lend Me Your Tongue!" Two Oceans Aquarium, January 2, 2018, https://www.aquarium.co.za/blog/entry/ks-kreature-feature-lend-me-your-tongue.
6. "Parasites," Seafood Health Facts, https://www.seafoodhealthfacts.org/seafood-safety/general-information-patients-and-consumers/seafood-safety-topics/parasites.
7. Tommy Leung, "The Crab-Castrating Parasite That Zombifies Its Prey," The Conversation, May 30, 2014, https://theconversation.com/the-crab-castrating-parasite-that-zombifies-its-prey-27200.

8. Katrina Lohan, "Marine Parasites: Crazy … and Really Cool!" Smithsonian, December 2012, https://ocean.si.edu/ocean-life/invertebrates/marine-parasites-crazyand-really-cool.
9. Christopher Bird, "These Shark-Eating Barnacles Will Give You Nightmares," Shark Devocean, October 23, 2015, https://sharkdevocean.wordpress.com/2015/10/23/shark-eating-barnacles/.
10. Allison Guy, "Celebrate Halloween with 3 Skin-Crawling Ocean Parasites," Oceana.org, October 27, 2016, https://oceana.org/blog/celebrate-halloween-3-skin-crawling-ocean-parasites.
11. Daniela Gomez, Jerri Bartholomew, and J. Oriol Sunyer, "Biology and Muscosal Immunity to Myxozoans," *Developmental & Comparative Immunology* 43 no. 2 (April 2014): pp. 243–256, https://www.ncbi.nlm.nih.gov/pmc/articles/PMC4216934/.
12. "What Is Whirling Disease?" Stop Aquatic Hitchhickers, http://stopaquatichitchhikers.org/hitchhikers/others-whirling-disease.
13. Rosemary Drisdelle, "Candiru—A 'Don't Pee in the Water' Horror Story Debunked," Decoded Science, June 30, 2013, https://www.decodedscience.org/candiru-a-dont-pee-in-the-water-horror-story-debunked/31635.
14. Dante Fenolio, *Life in the Dark: Illuminating Biodiversity in the Shadowy Haunts of Planet Earth* (Baltimore, MD: John Hopkins University Press, 2016), pp. 239–240, 245.
15. Ed Yong, "How This Fish Survives in a Sea Cucumber's Bum," *National Geographic*, May 10, 2016, https://www.nationalgeographic.com/science/phenomena/2016/05/10/how-this-fish-survives-in-a-sea-cucumbers-bum/.

Glossary

biodiversity The variety of life in an ecosystem.

contagious Able to spread easily.

decomposition The decay of dead organic matter.

ecosystem A community of living things that interact with one another and their environment.

ectoparasite A parasite that lives outside the body of its host.

endoparasite A parasite that lives inside the body of its host.

habitat The specific place a creature lives.

helminth A parasitic worm.

hemi-parasite A plant parasite that makes some of its own energy through photosynthesis and gets the rest from its host.

host An organism off which a parasite lives.

hygiene The practice of keeping oneself clean to encourage health.

parasitoid A parasite that kills its host.

protozoan A single-celled organism that eats organic matter.

spore A tiny seed-like organ a fungus uses to reproduce.

symbiotic relationship The interaction of different organisms that enable both to survive.

Further Reading

Books

Labrecque, Ellen. *Clean Water*. Ann Arbor, MI: Cherry Lake Publishing, 2017.

Montgomery, Heather. *Something Rotten: A Fresh Look at Roadkill*. New York, NY: Bloomsbury Children, 2018.

Olexa, Keith J. *The Gross Science of Lice and Other Parasites*. New York, NY: Rosen Central, 2018.

Stiefel, Chana. *Animal Zombies! And Other Bloodsucking Beasts, Creepy Creatures, and Real-Life Monsters*. Washington, DC: National Geographic Kids, 2018.

Websites

Medical News Today
www.medicalnewstoday.com/articles/220302.php
Read more information on different kinds of parasites.

National Geographic
news.nationalgeographic.com/news/2014/10/141031-zombies-parasites-animals-science-halloween/
Learn more about five parasites that take over their hosts' minds.

Science News for Students
sciencenewsforstudents.org/search?st=parasites
Read articles on parasites.

Index

A
Aphidius wasp, 17–18
Australian Christmas tree, 27

C
candiru, 36–37
chiggers, 9–10
Claviceps purpurea, 23–25
corpse flower, 28–30

E
ectoparasites, 5, 9, 10
E. muscae, 20–21
endoparasites, 5
eye flukes, 31–32

F
fleas, 5
follicle mites, 10

G
Glyptapanteles wasp, 18–20
guinea worms, 4–5

H
helminths, 12
hemi-parasites, 26
honeybees, 18
hookworm, 14

J
jewel wasps, 17

L
lice, 7

M
mistletoe, 25–27
mites, 9–10, 18
mushrooms, 11, 25

O
O. unilateralis, 21–22

P
parasitoids, 16, 17
pearlfish, 37–39

R
ringworm, 11–12

S
Sacculina, 33–35
stemsuckers, 27–28
strangleweed/dodder, 28

T
tapeworm, 12–13
tongue-eating louse, 33